Library of Congress C

Ruffault, Charlotte
  Animals undergrou
  by Graham Underh
  Translation of: La vie sous la terre.
  Includes index.
  Summary: Focuses on the natural history of animals that live underground.
  1. Soil fauna — Juvenile literature. 2. Burrowing animals — Juvenile literature. 3. Animals — Habitations — Juvenile literature. [1. Soil animals.] I. Underhill, Graham, ill. II. Title. III. Series: Young Discovery Library (Series); 3.
  QL110.R84 1988   591.56'4 dc19   87-34616
  ISBN 0-944589-03-0
  Printed and bound by L.E.G.O., Vicenza, Italy

Written by Charlotte Ruffault
Illustrated by Graham Underhill

Specialist Adviser:
Dr. Donald Bruning
New York Zoological Society

ISBN 0-944589-03-0
First U.S. Publication 1988 by
Young Discovery Library
217 Main St. • Ossining, NY 10562

©1985 by Editions Gallimard
Translated by Sarah Matthews
English text ©1987 by Moonlight Publishing Ltd.
Thanks to Aileen Buhl

YOUNG DISCOVERY LIBRARY

Animals Underground

Have you ever dug
a hole in the ground?

YOUNG DISCOVERY LIBRARY

On the top, the soil seems quite rough and stony. Look closely and you can see pebbles and twigs, old bits of leaves and lumps of earth. Imagine what it must be like to be a tiny insect clambering over all that!

As you dig down, the soil feels finer and more crumbly. It is not so dry, either. This fine, damp soil is just right for tunneling in or wriggling through. Burrow further down still, and the soil gets harder and more and more stony until you reach a layer of rock. There aren't any animals that could dig through that!

What if the soil wasn't there?

Think of the plant roots reaching down — where would they find their food? And what about the animals that eat the plants? And the minibeasts: the slugs, the wood lice and worms that find shelter in the damp, dark soil away from the sun?

Peat is a damp black mixture of ancient, rotted plants and trees. The Irish peat bogs are a paradise for the springtails.

Podzol, the kind of soil in some Central European forests, is covered with shrubs and little fir trees. Mites thrive there.

Earthworms love rendzina, a thin chalky soil. You can see the worms turned up in the furrows left by the tractor as it plows.

In Africa, the hot sun bakes and cracks the hard, red laterite soil, but termites still manage to build their towering nests.

What is soil?

There is not just one kind of soil. There are all sorts of different kinds, depending on where you are. The Earth is built up of layers of rock. Together they form the Earth's crust. These layers have been there for millions of years, and vary in hardness and color and shape according to what has happened above them, in the air or under the sea, or below them in the hot center of the Earth. Ever since rocks were first formed, wind and ice and rain have been wearing them away, forming boulders, pebbles and soil — all the different kinds of soil, from the dry sand of the desert to the soil you have in your garden.

Springtails

Mite

Earthworm

Termite

A handful of soil is full of life!

Pick up some soil in your hands. It seems still and silent, but wait a moment and look carefully. Can you see any tiny creatures working away inside it?

Where is a porcupine's home?

A porcupine digs a small home for itself in between the roots of trees, and makes it cozy by covering it with leaves which it carries to the hole in its mouth, two or three at a time.

What's round and brown and covered all over with prickles?

A porcupine curled up to sleep!

Porcupines are solitary creatures: they mark the boundaries of their territories with a strong-smelling saliva. Males and females attract each other with this strong smell too.

Porcupines sleep all day, curled up in a ball with their noses touching their tiny tails. No one would want to disturb such a prickly bundle!

At night, they eat. When evening falls the porcupines uncurl and go snuffling about the important business of finding dinner: insects, slugs, earthworms. When they've finished, they wipe their faces and clean their prickles.

What can you find in a spadeful of earth?

Look quickly before all the little creatures scurry away! None of the animals who make their homes here like too much bright light. In a moment they will all have hidden under the weeds. Can you see the ground beetle with his black wing cases? And the mole cricket hurriedly tunneling away into the ground? He's called a mole cricket because he uses his powerful front legs to dig, just like a mole. And there's an earwig, crawling up a stalk to begin his night's work, feeding himself on the soft insides of plants and fruit.

1. Slug
2. Ground beetle
3. Mole cricket
4. Earwig
5. Shrew
6. Millipede

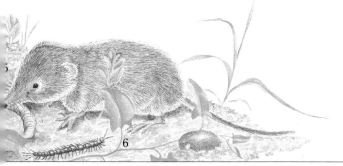

Underground Nurseries

As the days grow colder in the autumn, the snails who have fed on the green leaves all summer are slowing down. Soon they will settle down for a winter sleep. Before they do, the females dig tiny holes in the ground. They reach down these holes and lay their eggs at the bottom. The dark, damp nests will keep the eggs safe from hungry animals. In the spring, baby snails will hatch and make their way to the surface.

Baby snails have tiny transparent shells.

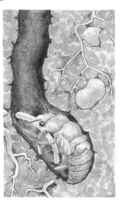

A cicada larva lives underground for four years.

A brief month in the open air

Insects go through three stages of development: first in the egg, then as larvae, and finally they **pupate** inside a cocoon and become an insect.
Beetle larvae live for three years in holes three feet under the earth. They eat their way through any roots that come their way until they get fat and sleepy. Then they weave a cocoon around themselves until they emerge as beetles. Once they have become insects they only live for a month.

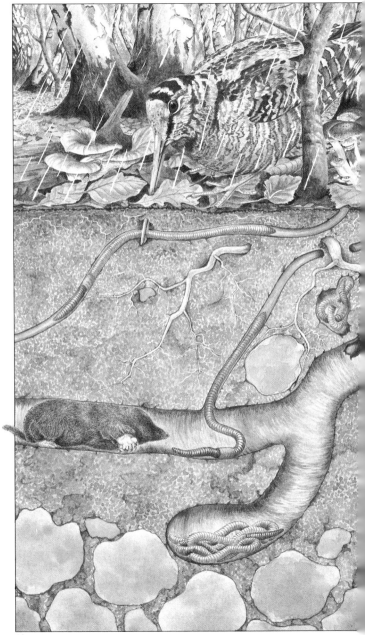

Earthworms eat a lot.
They get through the soil by eating their way through it! And as they're on the move all the time, they eat all the time. If you look at a lawn you might see little squiggly heaps of earth. These are the **worm casts** the worms leave behind them as they tunnel away. Worms are **hermaphrodite**: one end is male and the other female. Sometimes, if they are bitten in half, they may in time grow a new head or tail and wriggle away.

Earthworms are very precious.
The holes they bore let air and water into the soil. Never destroy worms.

Why is it the early bird that catches the worm? Because the damp night soil attracts the worms up to the surface, where birds like this woodcock can catch them.

A mole is not worried about 'rubbing its fur the wrong way'. Its hairs stand straight up and can be brushed comfortably in any direction.

Moles are tireless workers.

They crawl blindly along their tunnels, scooping the soil aside with their spade-like front paws and pressing it against the sides as they force their bodies through. Moles can't see very well, but they have a good sense of smell. Very sensitive hairs on their noses and on their tails help them find their way around, whether they're going backwards or forwards.

What are they doing down there undergound?

Moles eat worms that fall into their tunnels. They quickly bite the worms' heads and leave them paralyzed. The worms stay fresh and can be eaten later. Moles often build up larders of paralyzed worms.

Badgers are very clean animals. Every morning, they do their housework. In their dens underground, the floor has been covered with soft leaves and ferns. Badgers change the leaves everyday so that the den doesn't get dirty. They're very careful always to eat and go to the lavatory outside the den as well, so that it stays clean. Extra tunnels bring fresh air down to the main chambers.

How many rabbits are there in a burrow?

Ten, twenty, fifty, sometimes even more... In wintertime, all the rabbits come and go freely, but in the summer only the leader may sleep indoors. A guard on watch outside thumps on the ground with his back feet at the first sign of danger, then quickly runs away.

The only water rabbits drink is the dew that stands on the grass and plants which they eat. They spend all day nibbling away at leaves and roots with their long front teeth.

The desert may seem silent, but the fennec's huge ears help it to hear the tiniest sound.

How can animals survive in the desert?

During the day, the sun beats down, and there is very little shade. The only shelter is underground. The fennec, or desert fox, sleeps safe deep in his burrow, only coming out at night to hunt. Gerboa live in the desert, too. They make their burrows on the windy side of sand dunes. The wind blows sand across the entrance, hiding it.

Gerboa

The armadillo can curl into a ball, locking his scales shut with a little hook on the end of his tail.

The armored armadillo

Armadillos live in Central America. The nine-banded armadillo feeds in swamps and wetlands, digging through the mud to find worms. He can spend up to five minutes with his nose in the soil without breathing!

The Australian **duckbilled platypus** dives down to the bottom of a river and swallows mouthfuls of crabs, fish, weeds, and even small pebbles! He hasn't got any teeth, and the pebbles in his stomach help break up the food.

The platypus burrows at the river's edge. He's always careful to have several exits from his home.

Foxes don't like staying in one place.

Only the vixen digs a den to shelter her cubs.

What is a squatter?
Somebody who lives in somebody else's home without permission. Some animals are squatters. Weasels run up and down empty borrows, looking for the little rodents they like to eat. Their long, thin bodies are just right for tunnel hunting.

Foxes can be squatters, too.
They often move into old badger dens. Sometimes foxes kill more food than they can eat. Extra food is hidden to be eaten later.

Weasels often move house, but the homes they like best are moles' tunnels. Weasels spend all their time hunting, and eat almost their own body weight everyday.

In the autumn, in snowy regions. ermines' fur turns quite white, except for the tips of their tails, which stay black.

Prairie dogs are more like squirrels than dogs.

They have chisel-like front teeth for eating plants. Prairie dogs live together in vast underground tunnel cities. Sentries at the tunnel openings bark a warning if enemies come near.

When two prairie dogs meet, they give each other a little kiss to check that they know each other.

1. Rattlesnakes often follow prairie dogs right down into their burrows to catch and eat them.

2. Burrowing owls share the prairie dogs' burrows and eat the big beetles that scurry about in them.

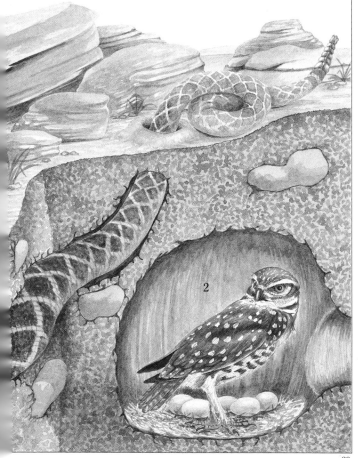

There are caves, too, underground.

Over millions of years, rain, falling on limestone rock, has worn channels, caves and grottoes. Animals live in these underground caves. They are called **troglodytes,** which means 'cave dwellers'. Hardly any light ever filters into these cold caverns, and the fish that swim in these waters have lost the use of their eyes. What good are eyes in the dark? They guide themselves by picking up vibrations in the water.

Flying visits

Bats often use caves like this to shelter in. They hang together in huge colonies on the walls, with the babies clinging to their mothers' tummies. The mother puts her wings around the baby, with just its head peeping out next to hers.

This lobster is white and blind, living away from the sunlight in the everlasting darkness of the cave.

Everyday, twice a day, the sea washes up the beach and then goes down again. This is called the tide.

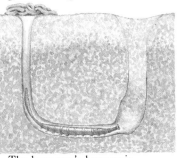

The lugworm's burrow is U-shaped.

At high tide, the razor clam siphons food into its mouth.

At low tide, the sand-colored weever fish leaves its fin sticking up through the sand. If you tread on it, it'll sting you!

Underneath the algae and seaweeds, there are thousands of tiny sand fleas. They can jump up to 72 times their own height.

seawater protects animals living along the tideline from
wind and the sun. It stops them getting too hot and dry.

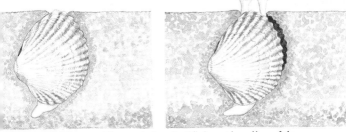

d is just within mouth's reach for shellfish as they lie safely
nored deep in the sand by their one powerful foot.

e transparent sand eels play in and out of the sand bars,
owing into the soft sand and then darting out again.

n the tide goes down, you can see the worm casts left by the
orms at the exits from their burrows.

Go and see for yourself!

Go for a walk through the woods on a summer's day, and see if you can find a **fox's den.** You can recognize it by the odds and ends of feathers, fur and bones at the entrance. But don't touch — the fox might get worried and never return. Then ask a grownup to take you back in the evening. Sit very quietly a little ways away, where the wind won't carry your scent down to the fox. Then wait, silently, silently...

A few peanuts and some honey will attract badgers.

On a grassy bank, the telltale holes of a rabbit burrow... You will be able to tell the nursery burrow by the little pile of earth at the entrance.

In thick woods, a hole in the ground... Fresh droppings in a hollow nearby and claw marks on the trunks of the trees: there are badgers here.

At the edge of a wood, a porcupine snuffles along his path. Don't forget – he'll roll up into a ball if you frighten him.

If you leave a saucer of milk out for a porcupine, he may come back the next night.

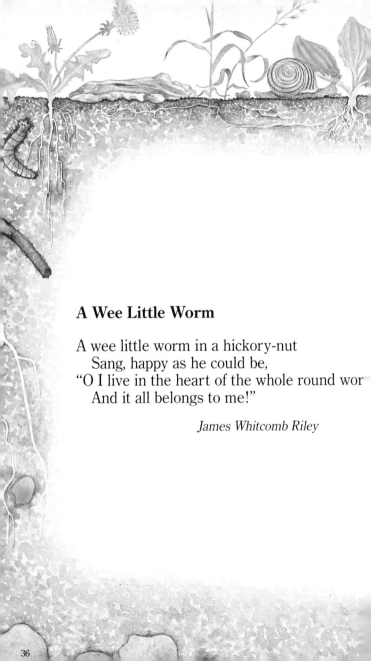

A Wee Little Worm

A wee little worm in a hickory-nut
  Sang, happy as he could be,
"O I live in the heart of the whole round wor
  And it all belongs to me!"

James Whitcomb Riley

Index

armadillos, 25
badgers, 20, 27, 35
bats, 31
beetles, 15
burrowing owls, 29
cicadas, 14
earthworms, 17, 19
earwigs, 13
ermines, 27
fennecs, 24
foxes, 26-27, 34
gerboa, 24
ground beetles, 13
lobsters, 31
lugworms, 32-33
millipedes, 13
minibeasts, 7

mole crickets, 13
moles, 19
platypus, 25
porcupines, 10-11, 35
prairie dogs, 28-29
rabbits, 22-23, 35
rattlesnakes, 29
razor clams, 32
sand eels, 33
shellfish, 33
shrews, 13
slugs, 13
snails, 14
soil, 7-9
troglodytes, 31
weasels, 27
weever fish, 32